AGRIMARKETING YOUR AGRIBUSINESS

AGRIMARKETING YOUR AGRIBUSINESS

A Guide To Marketing & Promoting Your Ag Business

LAURIE A. CERNY

AGRIMARKETING YOUR AGRIBUSINESS
A GUIDE TO MARKETING & PROMOTING YOUR AG BUSINESS

iUniverse books may be ordered through booksellers or by contacting:

iUniverse
1663 Liberty Drive
Bloomington, IN 47403
www.iuniverse.com
1-800-Authors (1-800-288-4677)

Because of the dynamic nature of the Internet, any web addresses or links contained in this book may have changed since publication and may no longer be valid. The views expressed in this work are solely those of the author and do not necessarily reflect the views of the publisher, and the publisher hereby disclaims any responsibility for them.

Any people depicted in stock imagery provided by Thinkstock are models, and such images are being used for illustrative purposes only.
Certain stock imagery © Thinkstock.

ISBN: 978-1-4917-9418-0 (sc)
ISBN: 978-1-4917-9419-7 (e)

Library of Congress Control Number: 2016905615

Print information available on the last page.

iUniverse rev. date: 04/19/2016

Dedicated to my late parents, Robert and LaVerne Cerny, and to the memory of their farm, Frosty Valley Farm.

TABLE OF CONTENTS

INTRODUCTION

I grew up on a farm in the 1970s. Our farm, Frosty Valley Farm, was located in what is known as the fruit belt of southwest Michigan. Over the years we raised everything from peaches, plums, and grapes to tomatoes, squash, and peppers. However, what our farm (along with my father's twin brother's farm next door) was known for was being one of the largest suppliers of fresh sweet corn on the Benton Harbor Fruit Market. Asparagus was our other big crop. Later we raised hay and soybeans.

I have also been a lifelong horse owner and an active competitor in various equine breed associations. We currently have a small horse farm where we also raise hay and herbs. I have my Dad to thank for what I know about farming—in particular, raising hay. I have 4-H and FFA (I was an active member of both all through middle and high school) for planting the seed in me to become a "communicator". I am still quite proud of the fact that I was the second female from my school's FFA chapter to become a "State Farmer". In addition, both my college education and 25+ years as a newspaper journalist helped to develop my writing and marketing skills.

The one thing I have observed over my years of involvement in the equine and ag industries is that for many business owners marketing and promotion are at the very bottom of their "to do" lists. The main reason is that most of us would rather be outside digging in the dirt then inside managing our businesses. While many can operate and still be profitable without doing much to market their agribusinesses, in today's competitive market and sluggish economy —doing nothing to keep the word out there about your business can be detrimental.

Coincidently, as I began writing this book, we were also in the process of going through things on our family farm (upon the death of both of our parents). As I cleaned my father's desk and file cabinets it was quite eye-opening to see (as I sorted through 20+ years of saved ag/farm related receipts, etc.) how many agribusinesses, which were still vital in the 1980s, are now gone. These ranged from canneries to farm equipment dealers and ag supply stores. It made me wonder: why? Was there no one to take over the business as the owners retired, or did the ups and downs of the economy over the past decade contribute to their demise? I suspect that for most they did not change with the times . . . some of which probably included a refusal to get onboard with computers and the Internet. Whereas in my parents' generation a lot of doing business in the ag industry was local—where you might have found a new vendor or supplier via word-of-mouth or in the Yellow pages—today many of us go to the Internet to find what we need. You can buy just about anything without ever leaving your farm.

So whether you have recently started your agribuisness, or have been running one for a long time, this book offers you a wealth of information to help with marketing and promotion.

A little bit of marketing and promotion can go a long way in attracting new customers to your agribusiness. Even in a small town where you may think everyone knows who you are and what products or services you have to offer, keeping your name out there is crucial to maintaining a strong customer base.

Good luck, and God bless!

The Ten Commandments of Marketing & Promoting Your AgriBusiness

1. Market your agribusiness

Marketing is the best way to get more business and to increase your potential customer base. Marketing includes everything from public relations and advertising to your business cards, flyers, brochures, websites, etc. It also means having the basic business tools like business cards, brochures, websites, signage, etc. A good marketing plan also includes promotional tools like baseball caps, T-shirts, pens, mugs, etc., with your business name, logo, etc. on them.

2. Determine your target market/audience

You must know your market in order to market to it. You also need to know who your market is in order to successfully provide a product or service. The main thing you need to determine is demographics. This, in simple terms, is your customer's make-up. Demographics generally include age, gender, income, ethnicity, etc. One way to find your target market is to survey current customers. A survey can be done in the form of a suggestion box at your place of business. Or, if you have a mailing list, which includes phone numbers and e-mails, you can contact customers either by phone, e-mail, or conventional mail.

3. Develop contacts with the media

The best way to get free publicity is to make a connection with the media in the ag industry as well as with mainstream publications. Sending press releases and/or making phone calls to a faceless person will yield far less results than having someone on the receiving end whom you have actually met, or at least talked to. Dealing with the media involves the similar kind of networking that you do as a businessperson. In addition, with mainstream media you need to make your contact comfortable with the terminology used by ag industry professionals and relative to your specific event or product/service. A good rule of thumb is to send as least two press releases each year, and to call two to three times. Call just to touch base with a reporter or editor and to ask if there's any interest in doing a story. Phone calls and/or a thank you card are appropriate follow-ups whenever a story is written. Publication of a press release (word for word as it was written) does not need a response.

4. Study your competition

In order to beat the competition, you must know the competition. This means more than just knowing of the other ag businesses in your area. You must also know what products and services the competition provides, as well as what they charge, and where they advertise.

Studying your competition takes a little research. One way, again, is to survey your current customers and find out where else they shop. Competition may also come from outside the traditional ag industry, so don't overlook this. Some big box (mainstream) stores also carry ag/farming supplies.

You also need to look at where the competition advertises - starting first with the phone book. Are they in the phone book as a business listing, or do they have something in the yellow pages, too? Cell phones have actually created a marketing void for many businesses especially ones that do not have landline phones. Unlike landline phone numbers mobile phone numbers are not automatically included in local phone books.

Besides the phone book it's helpful to find out if the competition has a website. If so, what does it look like and what does it include? Do they offer only information about their business, product, or service? Does the website also offer articles on topics farmers would be interested in and links to additional sources of information?

Next, you need to study the media. Does the competition advertise only in ag industry mediums? Do they use only regional and state publications, or do they also run national ads? What products do they advertise? Do they advertise with mainstream mediums including the local newspaper, radio stations, or television?

The main reason for studying the competition is to learn what they provide so you can provide it as well, and hopefully better. It also allows you to offer what they don't. If there are products or services competitors aren't providing, and it's something you can offer, it becomes a prime opportunity to fill a niche.

Knowing the competition, in combination with knowing your target market, can also help you to specialize.

5. Donate products, services, and your time

One of the best ways to get exposure for your ag product, service, or business is to donate your product or service, speak at a clinic or expo, or sponsor an event. Doing any of these things will come back to you ten-fold both in promoting your business and in giving back to the industry. In addition, the monetary equivalent of products/services donated are tax deductible.

6. Advertise

One of the first things businesses do when times are tough and sales are slow, is to reduce their advertising budget, or to cut it out altogether. Eliminating advertising is one of the worst things a business can do especially when sales are flat. When you don't advertise, existing customers forget about you and new customers don't know you exist.

When money is tight, there are things you can do to maintain your advertising presence. One way is to go smaller with ad size. For example, you could run a quarter page ad instead of a half-page or a full-page ad. You could also go from running a display ad every month to alternating between a display ad and a business card size (which many publications offer for a very reasonable rate). For example: you would run the display ad one month and the business card ad the next month. You could also run black and white ads instead of color ads.

Running a co-op ad is another option. Check with manufacturers of the products you carry to see if they are placing these ads. Usually the ads will list stores carrying their products (sometime there's a requirement of ordering a certain amount of their product). These ads will generally include your business's name, address, phone number, and website URL.

You should also investigate other advertising venues. Advertising in programs for expos and other ag related events and conferences can be very economical while reaching a very targeted audience.

7. Be honest about the level of service and the products you offer

Nothing hurts a business faster than not delivering what customers believe they should get. Farmers share information. Whenever they are unhappy with an agricultural product or service they will tell their neighbors and their farming peers.

8. Look for things that compliment your ag business inventory/services

Adding products and services can be a way to increase revenues and expand your customer base, as well as increase your exposure in the ag industry.

9. Be involved in the local community

Too often the thinking is that an ag businesses have nothing to do with the rest of the businesses in town. While farms and agricultural venues may not have a lot in common with other businesses, none-the-less they are still businesses and are subject to the same things any other businesses in town encounter.

First of all, being involved in the community from belonging to a local business organization or group, to periodically attending municipal meetings (village, city, township, etc.) is the best way to know what's going on around you and your agribusiness. Connections are connections and the contacts made with fellow business owners can lead to both referrals of customers who patronize their businesses, to support when potential changes in the community threaten the livelihood of your business.

Changes in ownership of property, development of land, changes in roads and highways, and changes in zoning ordinances are common things that will come up at local village, city, and township meetings. Going to these meetings a few times a year is a beneficial thing to do. If you aren't able to attend meetings, at least check a municipality's website for minutes from meetings. Reading the minutes will keep you somewhat informed, however, they can be vague on specific details, etc. As a business owner, you should also be aware of your township's land use plan. This is a plan, usually done by their engineer, that outlines projected land use for increments like 5, 10, 20, or more years. Many property owners have been shocked to see that their agriculturally zoned land is being planned for residential or commercial zoning in the future. The sad thing is that once a municipality approves a land use plan they work very hard to stick with it.

10. Make customer service a priority

Once lost, customer service is hard to regain. This is why customer service should always be a priority.

Mainstream businesses have operated for years under the adage that the customer is always right. It's probably the reason why many of the big box stores have become so successful. One of the attractions of these stores is that they are very easy to do business with. They have very convenient store hours, they take a wide array of credit cards, they have layaway programs, and they have reasonable return policies. In fact, most of these stores will take anything back if they sold it. You don't even need your original receipt as many stores can access your purchasing history through their store computers.

This isn't to say that anything goes, but providing good customer service is always a win-win situation. If a customer walks away from your store with nothing to gripe about, they won't complain to their friends. In fact, they might compliment your store and that in itself is very valuable marketing.

CHAPTER ONE

Basic Business Tools

Many agribusinesses have a common problem: they are good producing a crop or offering a service, but bad at being a business. Often there's a lack in the essential tools necessary to run a professional business. This being said—here are some guidelines for improving the business side of your agribusiness . . .

Run your agribusiness like a business: This means you need to establish set hours of operation and stick to them. The fastest way to lose customers is to not keep the hours you set, or to change them without notice. Even worse is to hang a sign on your door announcing you'll be back in a couple of hours, or closed for the day. This is both rude and unprofessional. Being consistent with the hours of operation you set and advertise is an essential marketing practice and a basic business tool. Similarly you need to have established *return* and *customer service* policies. Standing behind your products and services is another essential practice in marketing your business.

Purchase good office equipment: Any business in today's marketplace needs a good computer, Internet service provider, and a printer. Trying to look professional without the capability to type and print out letters and other correspondence, send email, and maintain a web page and/or social media accounts like Facebook and Twitter is impossible. And all of these tasks require a good computer. As for printers, buy the best you can afford. Doing so will pay off in the long run as your printed materials will look more crisp and professional. Purchasing one that also copies is worth its weight in gold as it will save you having to run to the office supply store or library to make copies.

Have a business phone/line: I still prefer a landline telephone system with a good answering machine or voice mail for a business office. Dedicating a phone line/cell phone number to your operation is essential to doing business with the public. This said, it is imperative that you also have a recorded voice mail message (or answering machine) that identifies the name of your business and hours of operation.

Often seasonal businesses will turn these things off during the winter. This is a big mistake—because callers will think they either have a wrong number or that you have gone out of business. Even if you don't plan to return calls during your off-season—at least keep your recorded message on—telling potential customers when you'll re-open and thanking them for their business during the previous season. Your phone message should also be updated throughout the year to let customers know when various crops are in-season, etc.

Cell phones, while wonderful, can still be inconsistent regarding service and the retrieval of messages. Having a fax machine is not a bad idea, although many businesses choose to email correspondence these days. In addition, buy a good calculator for use in both your daily business operations, as well as for when you sell to the public.

Invest in professionally printed letterhead, envelopes, business cards, and sales receipts: An attractive business card is invaluable. The first rule with business cards is to never make them yourself; they will always look homemade. Use a commercial printer—the more professional looking your card, the more professional your business looks to potential customers. With several online/mail-order printers (that do a great job for very affordable prices; please see the resources chapter in this book) there is no reason to think you can't have business cards printed. Make sure to include the name of your business, street address, phone number, and email address and website address (if you have a web page). If you are open to the public then you might want to include your hours, as well. Listing hours, however, will outdate your card if they change before you use all of your cards. And if you offer several products/services you might want to print these on the backside of your card; you never want to put too much on the front of a card as it will look cluttered.

Tips for creating a great business card:

- Keep it simple: include the basics (name, phone number(s), e-mail, web address, etc.
- Use a double-sided card for additional information (produce grown, products/services offered, etc.)
- Select a good paper stock
- Have a good contrast between card color and type color
- Print cards through an office supply store or online printer, rather than from your home computer/printer
- Keep photos and graphics to a minimum

What to include on your business card:

- Your business name: Include the full name of your business, however, if you have more than one business have a card for each one.
- Your business contact information: Make it easy for potential customers/clients to contact you. This means phone numbers, website address (if you have one) and e-mail address and cell phone number if you wish potential clients and customers to have round the clock access to you. If you are an operation where customers would physically patronize your business then include your street address, as well.
- Your business hours: For storefront businesses make sure to include days of operation and hours.
- Photos: Only use a photo if it is a great photo. In addition, it should only be used if it helps to tell the story of what you do.

What not to include on your business card:

- Coupons or discounts. Only offer a discount card or coupon if your card is a fold-over design. Design the card with a traditional business card on the topside and the coupon or discount offer on the backside card, which can be removed and returned for the offer.

- Contact information that may be obsolete in the near future. If you change e-mail addresses and cell phone numbers often you may not want to print them on your card.
- Too much information. Include only pertinent information like your business name, address, phone numbers, e-mail and website addresses.

Here are some additional tips for creating a great looking business card:

- Communicate the focus of your business. If your business name is your farm name, but you offer other ag services and products, you need to develop a statement objective that summarizes the scope of what you do. For example: Back Twenty Ranch (the business name) "We specialize in horses, hay & herbs" (statement objective). In addition, when you offer several services consider using a double-sided business card; use the front of the card for the name of your business and the backside of the card to list your services. Printing a double-sided card will cost more, but will allow more space and will help to keep the front of the card readable and uncluttered.
- Pick a good paper stock. The feel and look of paper speaks volumes about the business it represents. Choose a heavier stock. You can also pick a textured paper to give your card a rich feel.
- Pick good colors. Make sure there is a good contrast between the color of the card paper stock and the color of the type. Don't choose a dark paper and then use a dark color for type. Black type on a navy or midnight blue colored card would be difficult to read.
- Do something different. The majority of business cards are horizontal in layout. Try a vertical layout, instead.
- Select a standard font like Times New Roman, Garamond, and Century Schoolbook. All of these fonts are very legible. Italics, capital letters, and script type should be used sparingly.
- Consider embossed type. It costs more, but will give your card a more upscale look.
- If your business name is relatively short consider doing a vertical layout. A fold-over card is another option that will set your card

apart from others. It's also a great way to include additional information including services offered.

- Make sure that you check, double check, and then check one more time all of your information on the card. If possible, have someone else check your card before it goes to print. It is very easy to make mistakes—especially with phone numbers, e-mail addresses and street numbers.

Likewise, if your business requires you to write hard copy correspondence then you want to have a nice letterhead and envelopes. Again, include the pertinent information like your address, phone number, and email and website addresses. Forgo hours of operation—especially if they vary throughout the year. You can always include your hours of business in the body of your letter if necessary.

Professionally printed sales receipts/invoices—with the name of your business, etc., are also nice to have and are another means of advertising your business. Customers keep them for tax purposes and when they pull them out at tax time, they are reminded of your business.

CHAPTER TWO

Signage

Like business cards, your sign tells a lot about your ag operation. A run-down sign with peeling paint and words crossed out does not send a very professional message. Nor does a sign that has weeds or overgrown bushes surrounding it.

Your sign is the first thing customers see so make sure it's in an effective location, looks great, and meets local zoning requirements. If your sign doesn't look professional, or if it's missing letters and/or in need of a new paint job it may turn potential customers away. How often have you questioned the quality of service from an auto body shop, or a restaurant with a dilapidated sign out by the road?

For businesses located in extremely rural areas and on a backroad is it also important to have directional signs at major intersections. Remember: a lost customer is lost business. Always check the zoning ordinance in your community for regulations on signage, and never erect a sign in the right-of-way.

Sign location: The most likely place to locate your primary sign is in front of your agribusiness. For the most part, as long as your sign is within the size and material requirements outlined by your township's zoning/sign ordinance, and is not in the right-of-way, you can place it wherever you like. Obviously for a roadside stand it makes sense to place the sign either on the side(s) and/or on the front of the stand. If you have ample road frontage placing signs in advance of your location is also beneficial. Always make sure that signs are secure with adequate posts, etc. Never use telephone or other utility poles for your signs; these poles belong to that utility and they can legally remove and dispose of your sign. If you don't have road frontage that allows for signs in advance your business you will have to be more creative in driving oncoming traffic to your business. Getting the permission from a neighbor in either direction of your business (to locate a sign) is your best option. Offer some kind of payment for the use of their property—whether it's a small annual rent or compensation with some of your produce or products.

Directional Signs: For businesses located in extremely rural areas is it also important to have directional signs at major intersections near your business. Never erect these signs in the right-of-way or the open area at an intersection. You always want to periodically check these signs to make sure they are still directing people in the right direction.

Temporary Signs: Temporary signs like "sandwich" signs (which can be set out) and banners are a convenient and cost-effective way to market seasonal products or services, sales, and special events. Again, make sure they are allowed within your local zoning and make sure they are anchored well enough to withstand an unexpected wind storm.

Zoning Issues: Check with your village or township about ordinances pertaining to signs before putting up a new sign. Most have restrictions for the size of a sign and setback requirements from the road/right-of-way. Sometimes municipalities will also dictate the type of sign, including whether it can be lighted, has movable elements, and what materials it can be made out of. They may also have ordinances regarding temporary signs—usually outlining the length of time these signs can be used.

Vehicle Detailing: Detailing your vehicle with your agribusiness name, address, phone number etc. is essentially a movable sign that offers advertising wherever you go, and a great way to advertise when you attend the county fair and other ag events and expos. It, however, can be expensive to do. And it's also permanent. If you want something cheaper and less permanent, consider a magnetic sign for your car or truck. These are inexpensive and available through several online printers. You can easily put them on and take them off of your vehicle. Window decals and bumper stickers are another option.

CHAPTER THREE

Marketing Materials

Twenty years ago businesses needed a communication or marketing department, or at least needed to hire someone to create marketing materials like postcards, brochures, flyers, or newsletters. This is not the case today. Most anyone with a computer can create these things without too much effort. Online printers offer wonderful templates that will help you put together postcards and flyers that are both affordable and look professional. Purchasing a design software/print shop program will also allow you create good-looking flyers and newsletters. There are many affordable programs available for less than $50.

Overall Guidelines:

- Define the purpose of your promotional material
- Choose a layout complimentary for the medium you choose
- Balance white space (unused space) with text and graphics
- Select paper colors that correspond with business, product, or service. For agricultural based businesses earth tone colors are very complimentary.
- Pick a standard font like Times New Roman, Garamond, Century Schoolbook for text
- Text body should be done in 12 point. Headings 2-4 point larger
- Keep text style like "bold", "italics", etc. to a minimum
- Use capital letters sparingly (use for one or two words only)
- Use bullets for lists of services or products offered
- Use numbered lists for information with sequential order
- Edit your final draft at least three times

Postcards: I love postcards for two reasons: they are cheaper to send then letters and they are a great way to make announcements. They are a great way to let customers know about a sale, special event, location change, or change of business hours. They can also be used for a coupon. Even better is that today you don't have to print 500 or more of these to get a printer to run your order. Again, online printers will allow you to print as few as 25 at a time.

Tips for creating an effective postcard:

- Pick an attractive design for your front and keep the text minimal on the front side
- Organize your text on the backside in a logical manner and make sure you ask the receiver for some type of action like visiting your website, calling for a quote, or bringing the postcard in for a discount, etc.
- Make sure to leave ample room for an address, stamp, and at the bottom for scanning stickers the post office may apply.
- Include your return address if you want postcards returned in the case of an incorrect address.

Brochures: These are my least favorite to do as they take a lot of time to write and design. They are also more expensive to print and to mail then postcards. Before doing a brochure, make sure you have enough information to tastefully fill all of the panels of the brochure. There's nothing worse reading a brochure where nominal information is repeated to fill space.

Tips for creating a professional looking brochure:

- Have a clear purpose/message
- Organize information by the brochure panel
- Put contact information on front or back panel
- Don't mix too many background colors
- Avoid visual noise caused by using too many photos and graphics

- If a coupon, registration card, or something else that can be removed, is included make sure its backside does not include pertinent information
- Don't mix more than two font styles

Flyers: These can be created on most computers using Word. Other than printing or copying them—there is little overhead associated with making a flyer. To top it off, you can put them up for free any place in town where there is a bulletin board. How wonderful! Most ag/farm supply stores have such boards near their restrooms. Make sure to take either tape or a push pin/tack with you and stick several on the board so they can be taken. Some stores will want you to date them—as they take flyers down that have been up for more than 30 days.

Tips for designing flyers:

- Define the sole purpose of the flyer
- Include photos as attention getters
- Use a large enough font to view from a distance (14 point or larger)
- Use color paper if competing with numerous flyers
- Include phone # and other contact info on flyer
- Include pull tabs with contact info included on them as well as a reminder of what the number is for: "Tractor Parts & Service: 269-273-0347"

Newsletters: Newsletters are a great way to keep customers and potential clients informed about your products/services, as well as important industry news. It's a good medium for coupons and special offers. Moreover, if you include tips and "to do" lists and/or advice there's a good chance the issue will be kept or at least passed on to another potential customer.

Christmas Cards & Calendars: Christmas cards personalized with your business name and/or with a photograph of your employees is always a nice touch. To encourage patronage in the New Year put a coupon or discount offer in with the card. Calendars, although a little pricey to do, are a great way to keep your business name in front of your customers throughout the year. Don't be tempted to skimp and get the cheapest calendar possible. The nicer the calendar the better your chance that it will get used and not thrown out.

CHAPTER FOUR

Promotional Items

We all love to get free pens, t-shirts, and baseball caps. The beauty of these promotional items is that they are a marketing tool that keeps on giving. With low or no minimums being offered by many promotional product companies, even if you're on a tight budget, you should be able to afford one or two promotional items for your business.

Pens, Key Chains, Magnets: Pens with your business name and contact information are always a good promotion. If you aren't printing your logo on your pens you can easily find decent ones for less than 50 cents each (for 100). They are great to have at your counter when a customer signs their bill, possibly taking the pen with them. They're also a good item to give out at trade shows and expos. Magnets (usually containing the same information as your business card) are fairly affordable and make a useful item for your customers. The hope is that your magnet will end up on their refrigerator—where it will be seen every day. Key chains are another popular item, however, they may or may not end up being used once the customer returns home.

Mouse Pads, Coffee Mugs, Tumblers, Sports Bottles, Post-It Notes & Other Office Supplies: Anything that your customers can use in their day-to-day activities and at their job is an ideal way to keep the name of your agribusiness in front of them. These items can last and be used for years, therefore justifying the expense associated with printing them. Again, if you have a tight budget you will want to reserve giving these promotional items to your regular customers or to potential ones.

Hand Sanitizer, Lip Balm, Gum, Sunglasses: Personal care items (except for chewing gum) have decent longevity as promotional items.

They are a little more costly, however, there's a better chance that they won't be given away or thrown out. Gum or candy, with packaging that features you company's name/logo, is a welcomed and different promotional item. However, once consumed your promotion is generally gone. You can increase the staying power by printing a special/ discount on the packaging . . . something like "10% off your order with this wrapper". Sunglasses, with the name of your business printed on the sides, are a useful item and are also fairly affordable.

T-shirts & Hats: T-shirts and baseball caps are great promotional items that both advertise your agribusiness to whoever wears the shirt or hat, and to everyone who sees it. Again, these things are more expensive, but you can easily get them done for under $15 a shirt/hat. Some online printers will let you do singles (you'll pay more). Most others have a minimum of three shirts/hats. Obviously the more you print the cheaper you can get them for. If you are creative enough to design something with a catchy farm/ag related mantra or saying as the main artwork, and put your business name secondary . . . you might have a market for selling them at your business. This would enable you to print a higher quantity and to possibly make a little money from their sales. The rest of the time you can use t-shirts and hats as giveaways/gift with purchase and as "thank-you" tokens for your faithful customers. While it's nice to use a local printer, online suppliers (see resources chapter) offer free set-up for art and have hundreds of print styles and artwork available to use that will enable you to design a custom t-shirt yourself.

Tote Bags/Soft Sided Coolers/Drawstring Backpacks: These are great items that offer advertising for your business every time they're carried to an event, etc. So many people are using tote bags in place of shopping bags when grocery shopping; again making a grocery tote bag a good vehicle for advertising your agribusiness, products, and services in town. Drawstring backpacks are also affordable and a popular item with younger customers.

Office Clothing: T-shirts, polos, vests, etc. featuring the name of your business/logo are all great ways to create a professional and uniform look for your agribusiness. Not only are these clothing items great for promoting your business when customers patronize your office/farm stand . . . they become a walking billboard while attending ag industry events, or even when you and your employees run errands before and after work.

The Unusual: There is a lot to be said about coming up with something different to give out as a promotion. Currently among the unusual promotional items that can be printed inexpensively are poop bags (bags used for picking up your dog's doo/stored in a bone shaped container that can be hung from a keychain) and seed packs (with your business name printed on them). The neatest one I've seen is where the seeds are embedded into biodegradable paper that could be planted. At the high end of unusual promotional products is privately labeled wine/beer.

Employee Incentive Awards: Director/chairs and folding sports chairs, laptop/tablet covers, briefcases, garment bags, etc. are nice things to give your employees for various year-end awards. Not only are these items that get used, they are usually used when your employee is out in the public. Yes, they are a little expensive to do, but when used as an employee gift/incentive award the investment becomes two-fold.

CHAPTER FIVE

The Media

Developing a relationship with both your local media, as well as ag industry media, is highly beneficial in the promotion of your business. Whenever you are mentioned or featured in the local paper or in and an ag related tab or magazine it's free advertising for your ag products and services.

You always need to keep in mind that both print and broadcast mediums work on deadlines. Stories can either be done as "advances", which are in advance of the event (these usually run within the week of an event), or current coverage or breaking news (generally runs within 24 to 48 hours). Another type of story is a "feature"; this is the one we all hope for because they are usually longer stories and involve several photographs.

Many publications (especially magazines), television stations, and online publications have what is known in the publishing industry as "editorial calendars". It is always good to know what these are for the year so that you can pitch yourself or your business as a possible story for a topic they are covering a certain month. Let's say an ag industry magazine is devoting their April issue to "Sustainable Farming" —and your vegetable farm has been actively embracing sustainability practices for years. Obviously you would have a better chance of getting coverage in that issue then in an issue devoted to "Raising Organic Beef". If you are not able to locate an editorial calendar on the publication's general website, request a media kit . . . or a rate card. Since these items are geared toward potential advertisers the editorial calendar is often included so that ad clients can target market their ads per issue.

You also need to keep in mind the "lead time" that is associated with weekly and monthly publications. Weekly newspapers and farm tabs usually work about two weeks to ten days out from when they publish. This means if you hope they will do an advance story on your open house or event you should contact them two to three weeks before the event. Monthly magazines can work three to four months out from the issue publication date. Quarterly publications may work as far out as six months. Therefore, when you deal with magazines you need to make sure that you contact them well in advance of the issue you hope to appear in—this might mean as much as six months ahead of time.

The first step toward developing relationships with the media is doing a little research. You need to find out who the editors/reporters are for the sections of your local newspaper that may run ag related stories, as well as the TV reporters who regularly cover businesses and/or outdoor activities and events. Likewise, you will want to find out who the editors and writers are for ag industry magazines and newspapers that specialize in your particular niche. Once you have found these people, compile a media contact list that includes their name/position, and a working email address and phone number.

Reporting has long history as being a phone-based industry (although many reporters today do a lot of their work via email). Your first contact will likely be more successful if it's a phone call. You may have to leave a message, or may get routed to an operator or assistant, but that's OK. Local, daily newspapers all have deadlines (usually in the morning) as well as meetings . . . when these individuals obviously are not available. Find out these times and avoid calling then.

Once you have made contact with reporter or editor remember that they will usually be pressed for time—so try to keep your conversation short and to the point. Introduce yourself and quickly tell the reporter/editor who you are and a little about your agribusiness. Offer to be a source for any of their articles on things related to your farm or business. If you have an upcoming event or open house that you hope they will cover—let them know when it is. Better yet is to suggest story angles and photo opportunities. Anything that involves kids and animals makes a great photo opportunity, as does a live demonstrations like shaking cherries or cutting down Christmas trees.

Whenever you or your business gets coverage (and it's accurate) take time to thank the reporter. Often reporters and editors are only contacted by their sources when there's a mistake or they if they have a beef with the story. A thank you note letting them know they did a great job is always welcomed.

CHAPTER SIX

Writing Press Releases

Writing press releases is not for the faint of heart. They can be difficult even for the experienced public relations writer and are downright tedious for the novice. Besides phone queries with your media contacts, press releases are the main form of communication that helps to get coverage. Some mediums will run the releases word-for-word. Others use them as background information and will want to do an interview with you in order to write their own story.

The one thing you don't want to do is to send out press releases about nothing. Just because your farm stand is opening for the season doesn't make it reason to write a press release, or for story coverage. Always keep in mind the "so what" factor. Why would anyone care?

Things that the media like to cover include holiday events and activities, kids and animals, demonstrations, and the unusual. Let's say you've grown a 50-pound pumpkin or a carrot shaped like chicken. It's unusual, which is newsworthy. Likewise, if your event has a well-known politician or other notable person making an appearance, the press will probably cover it.

Tips for writing a press release:

- Keep it short—always keep press releases to one page
- Follow standard format (see sample) —including FOR IMMEDATE RELEASE: at the top, along with contact information and a focused subject line
- Always put a -30- or ### at the end of your release to indicate the end of the release
- Lead with the most interesting information

- Include possible photo and interview opportunities
- Include interesting quotes (if possible)
- Email your press release as a straight email (do not send an attachment)
- Include "Press Release for . . . (whatever it is for) in your subject line
- Postal mail releases only to a specific reporter/editor
- Follow-up with a phone call about a week after sending your press release

SAMPLE

FOR: IMMEDIATE RELEASE   Contact: American Agritourism Council
americanagritourismcouncil@aol.com 269-657-3842

2016 Midwest Agritourism Expo Offers Free Admission

The 2016 Midwest Agritourism Expo, March 10, 2016 at the Van Buren Conference Center in Lawrence, MI will be free for attendees. The expo, which runs from 2-6 p.m. will feature educational sessions on agritourism, liability, and marketing. In addition, vendors—from packaging suppliers to farm store products will be part of the expo's trade show.

According to Laurie Cerny, executive director of American Agritourism Council (AAC), anyone owning a farmstand, winery, u-pick produce operation, cider mill, corn maze, or pumpkin patch should attend. AAC is hosting this year's expo.

"Our goal for the expo is to offer both existing agritourism businesses and potential venues an afternoon of resources." She added, "Agritourism is one of the fastest growing areas within the ag industry in the United States. Adding a venue like u-pick, a bakery, a tour of your farm, or a seasonal event will significantly increase the revenue of a farm operation."

The general session on "Agritourism" will be a panel discussion led by Cerny and will cover things like "What is agritourism?", "What it's not?", "Popular venues to add", and more. Panelists include, Ron Goldy and Mark Longstroth, who are MSU Extension Educators, and Beth Hubbard, Michigan Agritourism Association president.

Mark Epple will cover the liability of allowing the public onto a farm. Epple, who graduated from MSU with a horticulture degree, is a Farm Bureau Insurance agent. He also operates a 100-acre fruit farm in southwest Michigan.

The last session of the afternoon will be "The Top Five Ways To Promote Your Agritourism Business."

Vendor tables, which are also free, are still available. For more information contact AAC at americanagritourismcouncil@aol or call 269-657-3842.

-30-

Finding The Focus . . . .

Answer the following whenever writing a press release then use the answers to write your lead sentence and subsequent paragraphs:

Who:

_____

_____

_____

What:

_____

_____

_____

Where:

_____

_____

_____

When:

_____

_____

_____

Why:

_____

_____

_____

How:

_____

_____

_____

Use the Inverted Pyramid format **to organize a press release . . .**

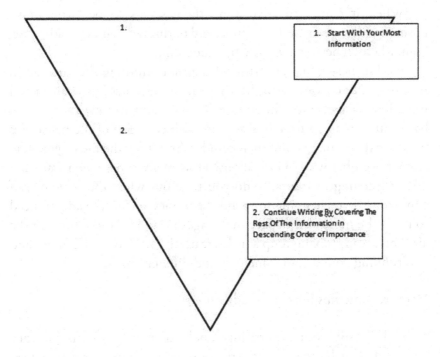

1.
1. Start With Your Most
   Information

2.
2. Continue Writing By Covering The
   Rest Of The Information in
   Descending Order of Importance

CHAPTER SEVEN

Advertising

When it comes to advertising there seems to be two camps: businesses that advertise all the time, and businesses that never advertise. Somewhere in between lies a happy medium.

For businesses that advertise all the time, running the same ad in the same publications or on the TV or radio becomes less of an ad and more like background white noise. Think about it: how many times have you seen an ad for a business or product in one of the magazines or newspapers you get and then couldn't find it for the life of you—let alone remember what it was about? How many times when watching TV or listening to the radio do you tune out when the same old ad comes on? If you're going to spend the money to run the ad you need to make the most of the space in a magazine and with the time on the air. This means mixing it up a little creatively with your ad content and also running your ads in some new and different media.

Here are some tips for regular advertisers:

- Diversify your advertising. Look for new mediums for your ads—including newsletters, websites, programs for conferences and sponsorship of ag related TV/radio programs.
- Think outside of the box in regards to your product. Viewers and readers remember the unusual and the memorable. Some of the best Super Bowl ads have been ones that have been extremely creative and that tug on our heartstrings.
- Don't hesitate to ask viewers/readers to act. Something as simple as hiding a clue in your ad and then holding a contest or drawing

for anyone who finds the clue in your ad gets readers/viewers to contact you.

- Think about creating an ongoing story with your ad content. In our town a local window manufacturer has created ads for his replacement windows that continue with a theme of a doctor/ scientist coming up with new features, etc. Years ago in the horse industry a distributor of a popular horse treat created a cartoon horse and each ad continued with an ongoing story using the character.

Here are a few guidelines for agri-businesses who have never or rarely advertise. These tips are geared toward getting the most out of your advertising budget and for maintaining an advertising plan even when finances are limited:

- Create an advertising plan. First and foremost request a media kit from the publications and other media you want to advertise in. A media kit will usually include rate cards, editorial calendars, ad sizes and a sample issue. Then, instead of relying on some hodgepodge plan for when you advertise, sit down with a calendar and plan what should be advertised when. Having a 12-month plan that corresponds with the editorial calendar is one of the most cost effective things you can do.
- Determine the best time to advertise. Some times of year are better to advertise than others and these depend on what you sell or do. Many ag industry publications also have issues with have higher print runs during months when they attend expos/ farm shows, etc. Running an ad in one of these issues will get you more exposure for the same price.
- Place less expensive ads. If you've been running four-color ads, and slow sales have limited your finances, consider going to black and white, or alternate between color one month and black and white, the next. You can also decrease the size of a display ad to save some money. Another economical means of advertising are the directories. Most publications have a directory where business card size ads appear. Classifieds are another economical way to advertise.
- Consider co-op advertising. Co-op ads are generally something offered by manufacturers to their vendors—where if you

purchase a certain amount of their product they include the name/address/phone number of your business in the ad for free.
- Offer to write a column and trade payment for advertising space, or trade for at least a mention of your business in a bio at the end of the piece.
- Ask for a discount. Most advertising mediums offer a frequency discount, usually starting when you commit to at least three ads. Some publications will let you mix and match ad sizes to get the discounts; others require that the ads qualifying for a frequency discount be a minimum size, or are all the same size. In addition, ask if discounts are available for pre-payment, cash payments, or for camera ready ads.

Other print mediums: Local Trade Lines, Shoppers, Auto/Tractor Dealer papers are very affordable mediums to advertise in. Most community Trade Lines/Shoppers offer very reasonable classifieds. Even display ads are quite cheap. Especially for agribusinesses that serve the local community like a farm equipment business, a u-pick apple orchard—these are great mediums to advertise in. They are generally published weekly and made available throughout the community for free. Auto/Tractor Dealer papers are a little more expensive, but are ideal publications for anyone advertising a farm equipment business.

Online advertising: Online advertising—especially Craigslist has pretty much eliminated classified advertising in most daily mainstream newspapers. The "Farm and Garden" area of the website is a favorite for anyone from horse and livestock owners to farmers. Also, anyone looking for fresh produce in their community might also search the site. It's free to put an ad on Craigslist—all you have to do is set up an account using a valid email address. Ads stay up on Craigslist for 30 days and then must be renewed. The downfall is that you may get responses from a bunch of quacks. Never use your business email as a way for viewers to contact you. Either give your phone number or use the contact link offered through Craigslist. Search engines like Google also offer ad packages (Google Ads) and are a good way to get exposure to a national market. Likewise, running an ad on a website that has something to do with your niche area of the ag industry can also be very effective.

Billboards: Billboards are expensive, but are an effective way to market your agribusiness to possible customers who are driving through your area. They are especially good for attracting the agritourist who may want to patronize local farm stands, u-picks, wineries, cider mills, etc. With billboards you want to make sure to pick a location where your target market will be traveling. So if you're trying to attract the seasonal summer vacationer to your winery, and there's a lake, amusement park, or other tourist attraction in the area—you need to advertise on a billboard along the road to your venue.

Placemats and coffee mugs at local restaurants: Don't rule these out. Running your ad on the paper placemats diner's use and on their coffee mugs is a great way to be seen by a local, captive audience. Recently, while out for breakfast in our town, I read all of the ads on both the placemat and on my coffee mug while we waited for our food.

CHAPTER EIGHT

The Internet & Social Media

It's nearly impossible to do any kind of business without utilizing the Internet and social media. Two givens for anyone with an agribusiness is to have an email address and to have a website. In addition, social media (Blogs, Twitter, Facebook, LinkedIn) are free and can really help to build your customer base. Additional social media include YouTube, Printerest, Instagram, which are very popular with a younger customer base (30s and younger) and work well for posting recipes, photos, projects, videos etc.

If using a computer is like speaking a foreign language then you need to get some help to set up a website or Facebook page. If you're on a limited budget try to find a family member or friend, who knows a little something about computer programing and setting up a website, to help. If you're in a college town you might be able to get assistance from a student studying website design/computer programing (who might need to do an internship/independent study for college credit). If none of these options are available—spend the money and hire someone to at least set up your initial website.

Email: If you have a business you need to have a valid email address. Email addresses are free and available with a variety of good providers like Gmail, AOL, and Yahoo. The rule here is to choose a professional address. Avoid cutesy things like farmergal@aol.com. If you are an agri-business that primarily does business with other businesses in the industry then stick with an address using your business name. If, however, your main customers are the general public—unless the name of your business easily identifies who you are like bakersblueberryfarm@gmail.com, you may want to try to come up with an address that includes something about your ag produce or service. Let's say you are a fruit tree supplier and for whatever reason your company name is Cortland Sales. This doesn't tell the email receiver anything about your business—especially for potential customers who have never done business with you. You can still include your business name in the address while adding "tree" to it for something like cortlandtreesales@ aol.com. Email is the most economical way to electronically market your business. It takes some time, but is worth the sweat equity. In addition to selecting an appropriate email address take the time to compile customer and potential customer email addresses and group them into a list in your "contacts" portion of your email provider program. Then, whenever you have a sale, bring in a new product or service, getting the word out is as simple as emailing the groups or lists in your contacts. Attaching a flyer or coupon is another means of distributing your information for free using email. You do want to be mindful of adding attachments with large photos and/or graphics because of the amount of time it takes to download the file from your email.

Websites: Many small businesses think they can't afford to have a website. This is a big misconception as there are some service providers who offer free web pages—the drawback being that they usually include their name somewhere in your URL address. On the other hand, with providers like GoDaddy and iPage it's easy and fairly affordable to have a professional looking website for as little as $10 a month. The nice thing about these providers is that you can get and purchase your domain name when buying a website hosting plan. This will enable you to get a domain name with a .com, .net, .org that nicely identifies you like "americanagritourismcouncil.org" or "agritrouristusa.com". GoDaddy and iPage hosting platforms enable you to set up, edit, and monitor your website from their site. In addition, they have very helpful technical support (provided free) available 24-7. Anyone who can use a keyboard and is comfortable using a basic word processing software like Word should be able to set up a website hosted by one of these two providers, or by a similar provider. As with your email address, your domain name can tell customers a lot about your business. Be careful when selecting it—since you register and pay for it for a minimum of one year, you want to make sure you are happy with the one you choose. Sometimes the domain name you would like is already taken—requiring you to be more creative, or to pick something else, altogether.

Blogs: If you like to write blogs are a great way to keep your information out there, as well as addressing ag industry issues, etc. There are several providers for blogs that are free—including Blogger (www.blogspot. com is a Google product). These are a good way to test the waters without spending any money; the downside is that they have a long URL address as they include the provider . . . like http://agritouristusa. blogspot.com. If you want a blog that has a URL with a .com, .org, net, etc. you will need to use a paid service/host provider. WordPress is a popular blog program offered by providers and is used by many large companies and well-known personalities.

Twitter/Facebook/LinkedIn: Free, free, free. This is the beauty of all three of these. Again, as with a blog, the more effort you put into these social mediums, the better tools for marketing your agribusiness they become. In a matter of weeks or months you can have hundreds, if not thousands, of followers. Twitter and LinkedIn tend to be better for business to business customer interaction. Facebook is great for staying connected with your end customers (the public). If you own a farm stand, winery, u-pick Facebook is a great way to share information about your crops, events, etc. The nice thing about Twitter is that you can do a search for any possible business or products, and if they have a Twitter account you can follow them. Facebook, however, is more restrictive and requests that you know the people you send "Friend" requests to. Totally different is LinkedIn—where you send requests to join groups. Someone with the group will review your profile before accepting you as a member.

Tips for using social media:

- When using social media for business marketing reasons, always try to look professional both with the user name you pick (try to get your business name) or something that identifies you with what you do in the ag industry. Likewise, use your logo or a photo of your products for your profile picture.

- Be selective of who you follow or friend. Check their profile, etc. to make sure that by adding them you are positively reflecting your agribusiness. Adding aspiring models and musicians may help to build your follower numbers, but won't do much to build your professional clout in whatever aspect of the ag industry you represent.

- Post on a regular basis. Daily tweets on Twitter and posts on Facebook will increase your followers. With LinkedIn join the conversations within the groups you have joined—this will help to get your name out in these groups.

- Never slander or defame anyone or their product/service. You could be liable for making such statements. Keep your gripes and complaints private and if necessary contract the individual either by phone, letter, or in person. Don't share it with the world—it will almost always come back to haunt you.

- Always check your spelling before sending anything out in a tweet or a post. Although you can delete your posts and tweets, it may take you a minute or two to do so and in that time hundreds of your followers may have read it, and even worse . . . they may have retweeted it.

CHAPTER NINE

Community & Ag Industry Involvement

If you do business with the general public and with the local ag industry you will find that being involved in the community is its own form of marketing. There are many ways to involve your agribusiness locally including:

Parades: It's always amazing to see how many people turn out for a parade—whether it's for Memorial Day, Fourth Of July, or a local school's homecoming. Being a participant in the parade lineup is a great way to market your agribusiness. Drive a tractor in it, use a horse-drawn wagon, or ride a Gator; really, you don't have to do much to get noticed. Add candy (to throw out to spectators) and you'll instantly draw a crowd. Make sure to attach a coupon or a business card.

Open Houses: Open houses are a nice way to bring new customers to your business. Offer a special discount, some food and drink, and even a guest speaker or demonstration and you will attract people. Always do a drawing so you can gather names and addresses of potential customers for your mailing list.

Schools & Scholarships: Supporting your local school and sports teams is a good way to promote your agribusiness. Doing so will get you recognized in programs for graduation and band/choir concerts, and at football games. Sponsoring school teams, clubs and events may also get you listed as a sponsor on placards around football and soccer fields, and on school t-shirts. Sponsoring a school scholarship for senior/graduates is also a way to get some publicity from the local media. Make sure you are at graduation to present the scholarship and be willing to talk to the press.

County Fairs: It goes without saying that if you own an agribusiness you should support your local county fair. You can do this both by sponsoring classes and/or grandstand entertainment, as well as purchasing livestock at the 4-H/exhibitor auction. All of these things will get you noticed by the fair audience and mentioned in local media coverage of the fair and of the auction.

Just as community involvement helps with marketing to the public, being involved in the ag industry on a local, regional, state, and national level is a great way to market your business to other ag business. It's also an opportunity to network and to get support and continuing ag related education.

Farm Bureau: The local county Farm Bureau is a good organization for any agribusiness to be a member of. Membership offers both local networking opportunities, as well as insurance support and discounts to other local businesses, etc. Farm Bureau also gives representation at the state level for legislative making decisions regarding agriculture.

Ag/Agribusiness Councils: If you have an interest in ag issues both at the state and national level, becoming involved in ag and/or agribusiness councils within your state can be beneficial. It's a great way to rub elbows with the movers and shakers within your state's ag industry.

Other Crop-Related Committees & Boards: Whenever you can belong to an organization that is specific to the crop you raise or the ag service you provide, you will reap marketing related outcomes. Serving on a board or committee is even better—because when the media need a source while writing a story that involves your service or crop, you will likely be asked for an interview. Serving as a source and being identified in an article, along with your agribusiness, is always free publicity for your business. What could be better?

CHAPTER TEN

Trade Shows, Expos & Conferences

Trade Shows: If you are a manufacturer of an ag industry product then attending a trade show is part of how you do business. Chances are most of these occur from December through March, and require some travel. To get the most from exhibiting at a trade show consider advertising in the schedule/program, speaking, or giving a demonstration. You may also want to do an advance postcard mailing to potential clients/ attendees to let them know your booth number and any specials you might offer during the expo. Trade shows are the perfect time to utilize your promotional materials as well as marketing materials (brochures, business cards, and product literature). Because you never want to run out of these things, keep a couple back in case you get low. Then you can at least make photo copies. Candy is often something exhibitors put out to attract visitors to their booths. Although candy will definitely attract people, they may not necessarily be potential customers. Save your money and put it toward promotional products that will at least help to get your name out there. Have bottled water and pop available to anyone who spends time in your booth inquiring about your product or service. As for your booth, make sure you have good signage—either in the form of a banner or table tenting. You definitely do not want to skimp on these. If you do: once the trade show gets busy it can be easy to get lost in the crowd.

Expos: These are generally agribusinesses/farmers to the end-consumer. They are very similar to trade shows except you are dealing with the general public. If your agribusiness is manufacturing food products or other consumables make sure you take samples. People love to try jellies, jams, breads, juices, etc.

With both trade shows and expos always make sure you know what will be provided. Will there be tables, chairs, and electrical outlets? Are you located in a corner, or is your booth on an aisle? Because vendor space at trade show and expos can be very expensive consider sharing a space with another business. Also, if you have a topic you can speak on contact management and offer to trade your speaking services for a discount on your booth.

Conferences: Conferences are usually educational in nature with attendees being active famers and owners of agribusinesses. Although you may be attending for informational reasons—a workshop, presentation or speech, training, etc., it is still an opportunity to market your agribusiness. Take business cards, brochures, pens, and wear a shirt or vest with your logo/business name on it. If the presenter asks for everyone to introduce themselves—use your introduction to let other attendees know what your agribusiness is all about.

CHAPTER ELEVEN

Resources

Business Cards, Letterhead, Postcards, etc.

Vista Print
www.Vistaprint.com
866-614-8002

123 Print
www.123print.com
800-877-5147

Colorful Images
www.colorfulimages.com
800-272-9209

T-Shirts

Rush Order Tees
www.RushOrderTees.com
800-620-1233

Jiffy Shirts
www.JiffyShirts.com/

Custom T
www.customink.com
800-293-4232

Promotional Products

Crestline
P.O. Box 2027
Lewiston, ME 04241
800-221-7797
www.crestline.com

4Imprint
877-446-7746
https://www.4imprint.com

Leaderpromos.com
855-521-8142
http://www.leader.promos.com

Positive Promotions
800-635-2666
http://www.positivepromotions.com

Empire Promos
877-477-6667
http://www.empirepromos.com/

Best Little Promo House In Texas, LLC
800-447-4440
http://promotionaslproducts.net

State Agriculture Organizations

Alaska
Alaska Division of Agriculture: dnr.alaska.gov 907-745-7200

Alabama
Alabama Agriculture & Industries: www.agi.alabama.com
Alabama Agribusiness Council: www.alagribusiness.org 800-642-7761

Arizona

Ag Advisory Council/Arizona Department of Agriculture:
https://agriculture.az.gov 602-542-4373
Agribusiness & Water Council of AZ:
https://agribusinessarizona.orhg 480-558-5301

Arkansas

Arkansas Agriculture Department:
http://aad.arkansas.gov/ 501-683-4851
Agricultural Council of Arkansas: http://agcouncil.net/ 501-376-0455

California

California Department of Food & Agriculture:
www.cdfa.cca.gov 916-654-0466
Agricultural Council of California: www.agcouncil.org 916-443-4887

Colorado

Colorado Department of Agriculture:
www.colorado.gov/ag 303-869-9000
Colorado Ag Council: www.coloradoagcouncil.net

Connecticut

Connecticut Department of Agriculture:
http://www.ct.gov/doag/ 860-713-2500
Connecticut Agriculture Information Council: ctaginfocouncil.org

Delaware

Delaware Department of Agriculture: dda.delaware.gov 302-698-4500

Florida

Florida Department of Agriculture & Consumer Services:
www.freshfromflorida.com 800-435-7352
Florida Agribusiness Council: www.agribusinesscouncil.org/florida.htm

Georgia

Georgia Department of Agriculture: www.agr.georgia.gov 770-535-5955
Georgia Agribusiness Council: www.ga.agribusiness.org

Hawaii
Hawaii Department of Agriculture: hdoa.hawaii.gov 808-973-9560

Idaho
Idaho State Department of Agriculture:
www.agri.idaho.gov 208-332-8500

Illinois
Illinois Department of Agriculture: www.agr.state.il.us 217-782-2172

Indiana
Indiana State Department of Agriculture:
http://www.in.gov/isda/ 317-232-8870
Agribusiness Council of Indiana: inagribuz.org 317-872-9812

Iowa
Iowa Department of Agriculture & Land Stewardship:
www.iowaagriculture.gov 515-281-5321
Agribusiness Association of Iowa: http://agribiz.org

Kansas
Kansas Department of Agriculture:
http://agriculture.ks.gov/ 785-564-6700, 785-296-3556
Kansas Agribusiness Retailers Association: https://www.ksaretailers.org/

Kentucky
Kentucky Department of Agriculture: www.kyagr.com 502-573-0450
Kentucky Agriculture Council: kyagcouncil.net

Louisiana
Louisiana Department of Agriculture & Forestry:
www.ldaf.state.la.us 866-927-2476
Ark-La-Tex Agricultural Council: www.altac.org 318-987-3769

Maine
Maine Department of Agriculture, Conservation & Forestry:
www.maine.gov.dacf 410-841-5700

Maryland
Maryland Department of Agriculture: mda.maryland.gov 410-841-5700

Massachusetts
Massachusetts Department of Agriculture Resources:
www.mass.gov/eea/agencies/agr/ 978-692-1904
Massachusetts Association of Agricultural Commissions:
www.massagcom.org/
The Center For Agriculture, Food & the Environment:
http://agumass.edu 413-545-4800

Michigan
Michigan Department of Agriculture & Rural Development:
michigan.gov/mda 800-292-3939
Michigan Ag Council: www.michiganagriculture.com

Minnesota
Minnesota Department of Agriculture:
www.mda.state/mn.us 651-201-6000
Minnesota Agrigrowth Council: agrigrowth.org 651-905-8900

Mississippi
Mississippi Department of Agriculture & Commerce:
www.mdac.state.ms.us 601-359-1100
Mississippi Agricultural Industry Council:
http://maicms.org/ 662-694-1895

Missouri
Missouri Department of Agriculture:
http://agriculture.mo.gov/ 573-751-4211
Missouri Agribusiness Association: http://mo-ag.com/

Montana
Montana Department of Agriculture: agr.mt.gov 406-444-3144

Nevada
Nevada Department of Agriculture: http://agri.nv.gov 775-353-3601

New Hampshire
New Hampshire Department of Agriculture, Markets & Food:
http:agriculture.nh.gov 603-271-3551

New Jersey
New Jersey Department of Agriculture:
www.state.nj.us/agriculture 609-292-5532

New Mexico
New Mexico Department of Agriculture:
http://daweb.nmsu.edu 575-646-3007
Farm To Table: www.farmtotablenm.org/ 505-473-1004

New York
New York State Department of Agriculture & Markets:
www.agriculture.ny.gov 518-457-8876

Nebraska
Nebraska Department of Agriculture:
www.nda.nebraska.gov 402-471-2341

North Carolina
North Carolina Department of Agriculture & Consumer Services:
www.ncagr.gov 919-707-3000

North Dakota
North Dakota Department of Agriculture: www.nd.gov.ndda 701-328-2231

Ohio
Ohio Department of Agriculture: www.agri.ohio.gov 614-728-6201
Ohio Agricultural Council: http://www.ohioagcouncil.org/ 614-794-8970

Oklahoma
Oklahoma Department of Agriculture, Food & Forestry:
www.oda.state.ok.us
Oklahoma Agricultural Cooperative Council:
http://okagcoop.org/ 405-241-5095

Oregon
Oregon Department of Agriculture:
http://www.oregon.gov/oda 503-986-4550
Agri-Business Council of Oregon: www.aglink.org 503-595-9121

Pennsylvania
Pennsylvania Department of Agriculture:
www.agriculture.pa.gov 717-787-4737
Penn State Ag Council: http://agsci.psu.edu/business/agcouncil

Rhode Island
Rhode Island Division of Agriculture:
http://www.dem.ri.gov/programs/bnatres/agricult/index.php 401-222-6800
Rhode Island Agricultural Council: http://riagcouncil.org/ 401-647-3570

South Carolina
South Carolina Department of Agriculture: http://agriculture.sc.gov/
South Carolina Agricultural Council: https://scagcouncil.wordpress.com

South Dakota
South Dakota Department of Agriculture:
http://sdda.ad.gov/farming-ranching-agribusiness/

Tennessee
Tennessee Department of Agriculture: https://agriculture.tn.gov 931-879-7917
Tennessee Agricultural Production Association: http://tapa.tennessee.edu/

Texas
Texas Department of Agriculture: www.texasagriculture.gov 512-463-7476
Texas Ag Council: www.txagcouncil.org
Texas Agricultural Cooperative Council: http://www.texas.coop/

Utah
Utah Department of Agriculture & Food: http://ag.utah.gov/ 801-538-7100

Vermont
Vermont Agency of Agriculture: http://agriculture.vermont.gov 802-828-2430

Virginia
Virginia Department of Agriculture & Consumer Services:
www.vdacs.virginia.gov 804-786-2042
Virginia Agribusiness Council: www.va-agribusiness.org 804-643-3555

Washington
Washington Department of Agriculture: http://agr.wa.gov 202-720-2791
Washington State Agribusiness Council:
www.agribusinesscouncil.org/washington.htm

West Virginia
West Virginia Department of Agriculture
www.agriculture.wv.gov 304-558-3550
West Virginia Agribusiness Council:
http://www.agribusinesscouncil.org/westvirginia.htm

Wisconsin
Wisconsin Department of Agriculture, Trade & Consumer Protection:
http://datcp.wi.gov/ 608-224-5012
Wisconsin Agribusiness Council: http://wisagri.com/ 608-437-4680

Wyoming
Wyoming Department of Agriculture: wyagric.state.wy.us 307-777-7321
Wyoming Agricultural Leadership Council: www.wylead.org 307-214-5080
Wyoming Business Council-Agribusiness Division:
www.wyomingbusiness.org/business/agri 307-777-6589

Trade Shows & Expos

Making It In Michigan: www.productcenter.msu.edu/miim

Farm Progress Show: http://farmprogressshow.com

Great Lakes Fruit, Vegetable & Farm Market Expo: 734-677-0503
http://www.glexpo.com

Illinois Specialty Crops Agritourism & Organic Conference: 309-557-2107 www.specialtygrowers.org

North Carolina Agritourism Networking Conference: www.ncagr.gov/markets/agritourism

OPGMA Congress: 614-884-1141 www.ohiovegetables.org

Indiana Horticulture Congress & Trade Show: www.inhortcongress.org

Fort Wayne Farm Show: http://tradexpos.com/fort-wayne-farm-show/

Women In Agriculture Conference: http://wia.unl.edu www.womeninagricultureconference.com

Midwest Agritourism Expo: www.americanagritourismcouncil.org

Ontario Fruit & Vegetable Convention: 905- 945-5363. www.ofvc.ca

Colorado Fruit & Vegetable Growers Association Annual Conference: 970- 667-4949, http://www.coloradoproduce.org

Michigan Farmers Market Conference: www.mifma.org. 517-432-3381.

The Agro Expo: http://www.theagroexpo.com/

Gulf Coast Agritourism & Ecotourism Business Development Conference: 850-623-3868 http://santarosa.if.as.ufl.edu

North Carolina Agritourism Networking Conference: www.ncagr.gov/markets/agritourism

National Farm Machinery Show: http://www.farmmachineryshow.org 502-367-5004

Spokane Ag Expo: https://greaterspokane.org/ag-expo/ 509-624-1393

World Ag Expo: http://worldagexpo.com/ 800-899-9186

Iowa Power Farm Show: http://iowapowershow.com/ 515-223-5119

Southern Farm Show: https://southernshows.com/sfs/ 704-376-6594

Western Farm Show: http://www.westernfarmshow.com/ 816-561-5323

Mid-South Farm & Gin Show:
http://farmandginshow.com/Exhibitors/News.asp

North American Farm Show:
http://tradexpos.com/north-american-farm-power-show/

Wisconsin Public Service Farm Show:
http://www.wisconsinpublicservice.com/business/show.aspx

National Agriculture & Agribusiness Organizations/Resources

http://www.fb.org/ (American Farm Bureau Federation)
http://www.usda.gov/wps/portal/usda/usdahome (United States Department of Agriculture)
http://nifa.usda.gov/ (Local Cooperative Extensions/Ag Universities)
http://www.agribusinesscouncil.org/ (The Agribusiness Council)

Agritourism

http://www.uvm.edu/vtagritourism/ (Vermont Agritourism Collaborative - University of Vermont)
http://www.ncagr.gov/markets/agritourism/index.htm
http://sfp.ucdavis.edu/agritourism/
http://www.naturalresources.msstate.edu/business/agritourism.asp (Mississippi State University's Natural Resource Enterprises)
http://www.eckertagrimarketing.com/eckert-agritourism-how-to-articles.shtml
http://naturetourism.tamu.edu/agritourism/facts-agritourism/
http://www.agmrc.org/commodities__products/agritourism/
www.nc-ana.com (North Carolina Agritourism Networking Association)

www.agritrouristusa.com
www.americanagritourismcouncil.org (American Agritourism Council)
www.michiganfarmfresh.com (Michigan Agri-tourism Association (MATA)

Industry Specific

http://agrinet.tamu.edu/groups/gfruitvg.htm Offers state listings of fruit/vegetable associations

Apples
U.S. Apple Association: www.usapple.org 703-442-8850

Blueberries
U.S. Highbush Blueberry Council www.blueberry.org
North America Blueberry Council www.nabcblues.org 916-983-2279

Cherries
National Cherry Growers & Industries Foundation:
http://maraschinocherries.org/about-ncgif/
Cherry Industry Administration Board: http://www.cherryboard.org/

Grapes
Winegrape Growers of America:
http://winegrapegrowersofamerica.org 800-241-1800

Potatoes
The Potato Association of America:
http://potatoassociation.org/ 207-581-3042
National Potato Council:
http://nationalpotatocouncil.org/ 202-682-9456

Corn
National Corn Growers Association: http://www.ncga.com/ 636-733-9004

Soybeans
American Soybean Association: https://soybeangrowers.com 314-576-1770

Hay
The National Hay Association: http://www.nationalhay.org/usfec

Wineries & Winemaking
http://www.winecountry.com
http://www.winecountry.com/
http://www.southwest-wine-guide.com/wine-tasting-etiquette.html
http://www.allamericanwineries.com/
http://extension.ucdavis.edu/unit/winemaking/
http://www.wineworkshop.net/
http://www.wineworkshop.net/

Bee Keeping
http://www.betterbee.com/Fun-Facts
http://www.beeworks.com/informationcentre/index.html
http://www.worldofbeekeeping.com/get-started/
http://www.dadant.com/

Fresh Produce & Farm Markets
www.unitedfresh.org
http://ffva.com (Florida Fruit & Vegetable Association)
www.ifvga.org (Iowa Fruit & Vegetable Association)
www.gfvga.org (Georgia Fruit & Vegetable Association)
www.msfruitandveg.com (Mississippi Fruit & Vegetable Association)
www.wisconsinfreshproduce.org
http://nfmaonline.org (National Farmer's Market Association)
http://mifma.org (Michigan Farmer's Market Association)
www.michiganagriculture.com (Michigan Ag Council)

Livestock & Equine
National Livestock Producers Association: www.nlpa.org 719-538-8843
American Horse Council: www.horsecouncil.org 202-296-4031

Ag Related Ad Papers

Fastline – Fastline.com 800-626-6409

TractorHouse – www.TractorHouse.com 800-307-5199

Ag Magazines/Newspapers

http://www.world-newspapers.com/farming.html - Website listing numerous ag industry publications
http://www.agriculture.com/
http://www.agrimarketing.com/s/101558

General Ag/Farming Publications

The Ag Magazine
530-279-2099 http://theagmagazine.com/

Successful Farming
http://www.agriculture.com/successful-farming

Farm Journal
http://www.agweb.com/farmjournal/ 877-692-4932

Fruit/Vegetable Publications

American Vegetable Grower/American Fruit Grower
37733 Euclid Ave. Willoughby, OH 44094. 440-942-2000
www.GrowingProduce.com

Good Fruit Grower
105S. 18th Street #217, Yakima, WA 98901 501-853-3520
www.goodfruit.com

Vegetable Growers News
75 Applewood Dr., Suite A, Sparta, MI 49345 616-887-9008
Vegetablegrowersnews.com

Fruit Growers News
75 Applewood Dr., Suite A, Sparta, MI 49345 616-887-9008
FruitGrowersNews.com

The Grower
355 Elmira Rd. North Unit 105, Guelph, Ontario NIK IS5 Canada
519-763-8728 http://www.thegrower.org/

Grain Publications

Feed & Grain Magazine
847-382-9187 http://www.feedandgrain.com/magazine/

Corn & Soybean Digest
7900 International Dr., Suite 650, Minneapolis, MN 55425 952-851-9329 http://cornandsoybeandigest.com/

American Soybean Magazine
12125 Woocrest Executive Drive, Suite 100, St. Louis, MO 63141
800-688-7692 https://soygrowers.com/ASAmag

Television and Radio Ag/Farming Shows & Programing

RFD-TV: www.rfdtv.com (Rural/Farming TV Cable Station)
http://tunein.com/radio/Agriculture-g233/ (website lists ag/farming radio shows in the U.S.)

Online Ag Related Media

AgriMarket.com www.agriculture.com www.agweb.com
www.FarmJournal.com
www.TopProducer-Online.com
www.BeefToday.com
www.AgDay.com

www.USFarmReport.com
www.ProFarmer.com
www.FarmJournalCornCollege.com
www.MachineryPete.com
www.Cattle-Exchange.com

Tractor Manufacturers

Case/IH
www.caseih.com
877-422-7344

John Deere
www.deere.com
866-993-3373

International Harvester/McCormick-Deering
www.mccormick-deering.com

New Holland Agriculture (Ford)
www.newholland.com

Massey Ferguson/AGCO
www.masseyferguson.us
877-525-4384

Tractor Parts

Yesterday's Tractor Company
www.yesterdaystractors.com
800-853-2651

Steiner – New Parts For Old Tractors
www.steinertractor.com
800-234-3280

Tractor Parts, Inc.
www.tractorpartsinc.com
270-678-9611

Cheap Tractor Parts
www.cheaptractorparts.com
270-382-2894

ABOUT THE AUTHOR

Laurie Cerny is an equine/ag marketing specialist. As a former newspaper journalist she has written over two thousand news and feature stories. They appeared in the Kalamazoo Gazette, Detroit News and Chicago Tribune. Many of the articles earned her state and national press awards.

Her book, Marketing and Promoting Your Horse Business is still in print and available from iUniverse and Amazon. She is a lifelong horse owner and farmer, and has been a featured speaker at several equine and ag expos.

Keep up with her on Twitter @ AgriMarketAgriBiz. You can also email her at laurie.cerny@aol.com. She can be reached at 269-657-3842.

The author and her Quarter Horse mare, Bees Molly Dolleo.

The author and her favorite old Ford tractor.

Printed in the United States
By Bookmasters